D1784066

BATTLEWORLD

BATTLEWORLD

KERRICK PATTERSON
SR Animation

AI Writers & Pub

To all the hearts who make a difference.
Perseverance
Faith
Hope
&
Love
- Yours truly
Sir Kerrick Dewayne Patterson

TRADEMARK

AI Writers & Publishing

NFT

Non Fungible Token - 1/10
Collection Owner - SR Animation

CONTENTS

Copyright   iv

Dedication   v

Trademark   vii

NFT   ix

1   Debriefing   1

2   Big Mission   3

3   Point Of Intrest   5

4   Happy Birthday   8

5   Check-In   10

6   Big Breakfast   12

7   AI1   14

8   Shipping & Handling   16

9   Toe to Toe   18

10   Defcon   20

11   Fire in the Hole   22

12 Great Escape   24

13 Home Base   26

14 War   27

About The Author   29
Signature   31
DocuSign   33
DocuSign   35
DocuSign   37
DocuSign   39

CHAPTER 1

Debriefing

Location: Qubar
Year: 2053

[machine opens] "where am I?" said Bruce.
"The cryostasis department," General Hammond said.
"Huh?" asked Bruce.
"We recently reactivated you welcome back solider," said general Hammond. Debriefing in fifteen minutes. Catch your breath and meet us in the Strat room," said general Hammond.
As you all know, we're planning to take down the ultra-bots' last known stronghold. recently we awakened two of our top covert operatives to assist- [door opens]
"Almost got lost," said Bruce.
"Speaking of the devil, welcome in have a seat Bruce," General Hammond said, raising his hand. A few things have changed since you've been gone. As we all know, this has

been a long fight. In the last nineteen years, we've taken down every last Ultra-bot stronghold. "Brought Bruce here back for the last one," General Hammond said, while chuckling. Ten Thousand square foot compound, Seven entry points and over five hundred civilian workers. They've logged every known entity in the world, so we brought you in for recon. Agent two one five, six six dash eleven will be joining you, codename Sarah. We want you guys to go in, identify any weak points and report back to us. "Think you can handle the job" he asked.

"Sir yes sir!" yelled Bruce.

CHAPTER 2

Big Mission

Location: Hunnington Place

Bruce knocked at the front door. "Its me. Open up!"

"man you look like shit. Come in," Kyle said, then closed the door.

"thanks man, it's hot out there," said Bruce

"ten years flew by quick. How are you buddy?" asked kyle?

"Ok, I guess it's my first time unthawing," said Bruce. "Stop by to see if you know anything about Fairfax?"

"No, but i heard it's full of high-roller recons like yourself," said Kyle. They filled the place up after you went under. Why, they send you on another death mission?" asked Kyle.

"Um, I hope not," Bruce said, while smirking. "Hey I thought you came to hang out?" asked Kyle. You guys never stop working. "I'm starting to believe you're exactly what you're fighting, a bot!" Kyle said, while standing by the robot figurine.

Kyle, don't be ridiculous. Bruce opens the closet door. I'm here to pick up my bag strictly protocol.

"Yea right protocol." said Kyle. "Ten years ago we'd hit the bar and grab us a few girls, this must be a Big mission?" asked Kyle. "Ten years wait, and a man doesn't wanna have any fun, what gives?" asked Kyle. "Hot date i dont know about?" asked Kyle.

"No, Kyle, i just don't want to screw this one up. I promise we can have that beer when I get back," said Bruce.

"Ok, I'll take your word for it. What's her name anyway?" asked Kyle. "How'd you- Sarah Polinski." said Bruce.

CHAPTER 3

Point Of Intrest

Location: Fairfax

"Fairfax city!" Bruce shouted. "Long time buddy. Seems like yesterday was basic training," said Bruce. The new re-model really changed this place a lot. Wonder if they kept the pizza stand?" asked Bruce. "Only one way to find out." Bruce said, reaching for his key card.

"Damn my key cards not working," said Bruce. Can you buzz me in?" asked Bruce. Agent 10353-21 Bruce Landry.

"That you Bruce?" asked Sargent Harris. Welcome back, come on in. "Sarah's in the lobby waiting for you," said Sargent Harris.

"Thank god, thought I was going to have to do a Facial Recognition search," said Bruce.

"Nope, not like old times were pretty up to speed now," said Sargent Harris. "you guys headed for bergall better be careful, they don't like visitors over there," Sargent Harris said, while walking away.

"got it, thanks" said Bruce. The Lobby's awfully crowded. Maybe I can still spot her. "should be easy," Bruce said, as he looks around. Dirty blond bingo.

"Hi are you agent 21566-11 Sarah Pelinsky?" asked Bruce. "I am she," said Sarah. "You must be the infamous Bruce Landry." Mind joining me for lunch? It's on me?" asked Sarah.

"Sure I'm starving" said Bruce.

"Is pizza Cool with you?" asked Sarah.

"Deal," Bruce said, walking towards the pizza stand.

"Two slices New York style!" Sarah shouted. "Make that three slices big fellas hungry," said Sarah.

"One pep one cheese please," said Bruce. "So how's Fairfax been treating you?"

"Well five Star Hilton, golden Corral, Pizza hut, Shooting Range, Indoor heated pool, and state-of-the-art gym," said Sarah. Cant complain, the remodel came just in time for us. How was Cryo?" asked Sarah.

Didn't feel a pinch, it was over in the blink of an eye. said Bruce. Pizzas ready. Let's sit down and go over mission details. Bruce said, pulling a chair up.

"Right, hows the general?" asked Sarah.

"Ok" said Bruce. "Quabar?" asked Sarah. "Hellhole" said Bruce. "still the best pizza in the world" Bruce said, slightly caughing

"Yeah Pretty much," said Sarah.

"Soo mission objective, we're being dispatched to Bergall City," said Bruce. It's the last ultra-bot stronghold. This is a Recon only mission. Gather intel report back to Quabar. I have ordered us not to engage, under any circumstances. Being captured is a high possibility. They've arranged a meeting with an inside man, a factory worker goes by the name of Cody. He'll be our guide around the wearhouse and compound. This should be an in and out mission. Person of Intrest Tetron and Koyballion. "Questions?" asked Bruce.

"No sounds like a piece of cake, let's do it!" Yelled Sarah.

"Well, let's load up and move out, transports outside waiting," said Bruce.

"Ok one, stop on the way to Bergall," said Sarah. I have to pick up some things from my old roommate Brittany in Hampton oaks.

CHAPTER 4

Happy Birthday

Location: Hampton Oaks, Ca

"The driveways full must be a party here or something tonight," Hey Sarah said, while walking towards Brittany. "Brittany Carrington, what did you do to your hair?" asked Sarah.

"Hey girl, I never told you I was going to join the Recon team back in April?" Brittany said. "No," replied Sarah.

Brittany asked who the jockey was, after mentioning that I had cut it two weeks ago.. "Who's the jockey with you?" asked Brittany.

"Bruce, this is Brittany, Brittany this is Bruce," said Sarah.

"Hi how are you?" asked Bruce. "Hey, I'm fine" replied

Brittany.

"So who's birthday?" Sarah asked, while grabbing a balloon.

"My cousin, she's turning seven. Figured I'd use my place this year for her party," said Brittany. "You here looking for your Shellwear?" asked Brittany.

"Yes! tell me you found it," replied Sarah.

"It's in the kitchen bottom righthand drawer," said Brittany.

"Thanks Brittany, you're a lifesaver. We're actually headed to Bergall for a light recon mission," Sarah said, while closing the drawers. If I need anything else ill call you.

"ok" replied Brittany

"got it," Sarah said, going through her bag. Everythings here, i just can't find my Facerec glasses. Damn, well, we're out of here. Thanks again Brittany.

"No problem girl" Brittany said, while walking her outside.

"Lightspeed" whispered Brittany.

"Lightspeed" replied Bruce.

"Lightspeed" said sarah.

CHAPTER 5

Check-In

Location: Bergall City

"Finally, we're here Bergall City" said Bruce. let's stay close, meet our contact and see how things work out here.

"Got it," replied Sarah, opening her iPad.

"First, let's check into our room. Heard they have the best stays in the galaxy," said Bruce. I know Scrapcom paid top dollar for this watch.

"Hi how are you? Welcome," said the manager. "hi how may I help you?" he asked

"Reservation number three ninety confirmation number one X six, seven three nine one one nine, zero," said Bruce.

"Ah yes, room nine zero one is ready for you," said the

manager. Here's your QR code.

"thanks" replied Bruce

"Airlavators open, let's go," said Bruce

"ninth floor," Sarah said, pressing the airlirlavator.

"woah! presidential sweet floor nice," said Bruce.

   "Yeah, what's the name of this place, by the way?" asked Sarah.

"lithoroma Stays," replied Bruce

"Six Stars definitely" Sarah said, slightly swiping the keycard.

"Yup, definitely wake me up in the morning for lite recon," said Bruce.

"You got it boss, I'm going to stay up a little while longer," said Sarah.

CHAPTER 6

Big Breakfast

Location: Bergall City

"Good morning sleepy head," said Sarah. Breakfast from waffle House.

"thanks its six am. I better get up," said Bruce.
"Yeah city recon this morning and meeting with contact named Cody" said Sarah. Got our load out bags ready, air tags and eye cameras. Equipment will be ready for deployment at entry points. "I've got us covered today," said Sarah. I'll be in the parking deck mapping coordinates.
"Sure thing, I'm right behind you" said Bruce. Hygiene, breakfast, Clothes and now let's find my keys and head out.
"Good room service," Bruce said, walking to the parking deck.
"sorry it took me so long. Great breakfast," said Bruce.

Yeah, room service, saw you were sleeping like a baby. Said Sarah.

Let's go. Cody should be at tropics by now. Bruce said.

"tropics and step on it!" Sarah said, yelling at the driver.

"roger that," the driver replied.

"In route to meet contact standby," Bruce said, over the Stratlink intercom.

Park around back. Let's be discrete he said.

"Are you guys with Strat Recon?" Cody asked. "Yes, come with us!" Bruce replied.

"No way you guys are hot buddy, let stay here" Cody said. First off, up, nobody has ever dared try to infiltrate their headquarters to this day. Soo they wake up, two nerdy recons to pull off a suicide mission. You guys sure have got some balls. The ultra-bots have a system, they find everything out and they see everyone. So if you think I'm going to just walk you in the back door and let you destroy everything, you got another thing coming. They only paid me to show your team around, not get myself caught up and possibly killed. "No way man wearhouse, then I'm out," he said. I take wire transfer cash or credit. "now let's go!" Cody yelled, and jumped into his vehicle.

CHAPTER 7

AI1

Location: Bergall City

"I'm feeling thirsty, so could you please stop here?" asked Cody to the driver.

If I am not mistaken, this is the new AI1 Station, is that correct?" asked Bruce, while pointing at the gas pump.

"Yeah, I helped build this one," replied Cody

"we just got one of these back home," said Bruce, state-of-the art, he mumbled. "Had to cost a million dollars to get one of these up," said Sarah. "yea give or take.. be right back," Cody insisted. Oh, my god! Yelled Sarah. "Bruce, there's a giant pink bird behind you," she said. "Woah!" Bruce said. "In the interest of safety, I request that you stand down and terminate all actions," Said Zeda. I am not your enemy. I'm only here to deliver a message. She said. My leader watches you from Mt. Saniah. He wishes you to join him. Decline and suffer the fate of this world, she said. "Wait a minute, who's your leader, and I thought ultra bots ran this, where'd you come from?" Asked Bruce. My name is Zeda, the leader of the super women. We survived WWIII by watching the humans and bots. The few of us that weren't killed off by the plague. She informed us that Bergall is a threat and is currently under surveillance for safety purposes. "yeah, tell me about it. Looks like everybody beat us to the punch," said Bruce. "Weapons of Mass destruction are being hidden here and the ultra-bots plan on using them against us," said Zeda. Our only hope is to form an alliance with humans. Or risk the planet being destroyed. Also, watch the one who guides you. He shall betray you as well. Be careful when you return to Qubar. "They enslaved you there and wanted you to die," she said. I will log this conversation as a formal greeting. Fair well! she yelled as her wings flapped in the distance.

CHAPTER 8

Shipping & Handling

Location: Bergall City

You guys look like you saw a ghost. Cody said, while closing the car door. Did one of the ultra bots pass by while I was gone? He asked.

"no sorry puzzled by the AI1 station, that's all. Looked up the schematics, it's almost two million dollars crammed into one thousand square feet," said Sarah. amazing, what's it powering you know?" She asked. No, I just work man, I don't ask too many questions, Cody replied. Let's go, the warehouse should be empty so I can show you guys around. "Let's ride," said Bruce. "park in the garage. I'll open it." Cody said to the convoy driver. "come on you guys," He insisted. This is the dock where we load up the products. One of the ultra-bots patrols this outside entrance. We package anything from tissue to firearms, whatever is on the demand list. break office in the back seats twenty. over here, we put up an inner room for heavy machine drivers. "Like a street?" Asked Bruce. "Yeah, like a little complex," Cody replied. This way, He said. The big vault here in the back holds any bio chemical weaponry in the world, this baby's blast proof and encrypted by not only one but five layers of quantum keys. last but not least, front parking lot. parks twenty-five cars and ten semi trucks as well. With state-of-the-art surveillance small private shifts, we control the production flow. "nice! Bruce whispered, walking around the parking deck. "Ok fellas, that's enough," said Sarah. Let's go. I got a bad feeling, she said.

CHAPTER 9

Toe to Toe

Location: bergall city

"Tetron I felt that!" E-foo yelled as he was closing a portal he arrived in. "Enough Let it go" he said. "E-foo, does your death announce my doorstep?" asked Tetron. Ultra-bots have taken over this planet's core. The dead humans walk, their heroes are gone, and the system is failing. We left all that was weak. Kept only those who were mighty warriors of the future! Tetron argued while pacing the floor, arms folded behind his back. For centuries, I have idly sat by and watched our races fight. now is the time we ultra-bots rise and take back what's rightfully ours. Earth's core Energy is enough to sustain our future endeavors. With you out of the way, I can finally move forward. You dare attack me for the girl. She invaded Bergall after we made a pact to stay away! yelled Tetron. She deserves to die along with the others. "So to what pleasure do I owe the gods for delivering you into my hands" he asked? You morale less coward, Ill kill you! E-foo yelled while throwing the first blow, which failed to land." ahh I'm striking him pretty fast, but I still can't land a blow E-foo said, thinking to himself. "you've gotten stronger, but your technique doesn't phase me. One stroke from my blade and you'll die! Screamed Tetron as he swung his weapon towards E-foo.

" You're nothing but a piece of metal," E-foo said. Screw you, you'll never get the orbs," he said. Oh, but yet I have already acquired them. said Tetron. "Now die!" Tetron yelled as the first swing landed. I told you, you idiot. "Now suffer your fate," he said. "The one will stop you, I promise.." E-foo mumbled as he was struck to the ground. "I figured you'd coward out," said Tetron. "Noo don't!" screamed E-foo while trying to shield himself. "ha the crushing blow" Tetron said. You should have been prepared, young lad. Next time stay out of bergall. ha ha ha ha ha. "The one will stop you!" E-foo said as he mumbled his last words.

CHAPTER 10

Defcon

Location: Bergall City

"mega-sphere, report to headquarters over! Tetron repeated over the voice-com. "Mega-sphere here over," he said. "Is everything alright, boss?" Asked Mega-sphere.

"I'm fine," said Tetron. I just got breached at headquarters," he said. Looks like E-foo and the humans were watching us the whole time. He said they've planned to infiltrate and destroy our headquarters.. They must have known we were going for the girl, too. They know about the orbs as well. "I want you to lockdown the entire city," said Tetron. Nothing moves. "In or out, you got it," repeated tetron. "I'll get right to it," said mega-sphere. "Do a Facial Recognition scan of Bergall's inner corridors. See if we have any unidentified persons, will you?" Asked Tetron. If you spot anything, do not engage. He said. Just report it immediately. "I will handle the rest," Said Tetron. "Copy that," mega-sphere replied. "Copy that boss," He said.

CHAPTER 11

Fire in the Hole

Location: Bergall City

"Bergall reeks of contamination," Said Maximus.

Show yourself. "You fool, Tetron will have your head for coming here!" Yelled Koy-Ballion." let me handle that, you all will pay for the wars you've caused," said maximus. You've discussed my father with your growing detest. "If you're the one they say will defeat tetron, then why do you come here to attack me?" Asked Koy-Ballion. "Life is yet a chess game," said Maximus, as he grabbed Koy-Ballion. Only the foolish ponder. Eternal slumber shall humble you in the afterlife. I'm done wrestling with you. Watch me crush your neck with my foot! Maximus said, stomping on Koy-Ballion's chest. "I shall squash you like a bug, He said. "you can kill me but the ultra bots will live on," Koy-Ballion said, struggling. We have already created a portal back, you're too late maximus. "you're too late!" He yelled. "Initiate a self detonation protocol," said Koy-Ballion. "you idiot," said maximus, as he ran for cover. "Dammit, I'm severely wounded. How am I going to patch up this one?" said maximus. He caught me off guard. "Who's that?" He asked. It's me Zeda. "I'm here to help you," Said Zeda. "Finally thank you. "Any word from E-foo?" Said maximus. "No Tetron doesn't spare anybody," she said. 'Why would he attack?' Asked Zeda. Maybe he saw something we didn't," she said "mabe," Maximus said crawling out of cover. "you'll feel better shortly, here take this for energy," said Zeda.

CHAPTER 12

Great Escape

Location: Bergall City

You hear those sirens!" Cody yelled, checking the computer. That's the sound of a lockdown. "something about to go down here," He said. "Lockdown?" Asked Bruce. "How much time do we have to get out of this place?" He asked. "I'm guessing nine minutes at best," said Cody. Until all seven doors close. You'll be stuck until the lockdowns lifted. the inner gates should come up any time now. "I have to go. Good luck!" Cody yelled as he's running off. "wait so you're going to leave us? Asked Sarah. "Asshole no wonder. I can figure this one out. Let's go Bruce," She said. "I gotta hurry, the micro basement in gate six will work," Said Cody. Glad I grabbed my key card. "Open inner wall gate six," Cody repeated to the system. "Facial Recognition scan complete, access granted.," The system replied. Welcome back to sector six Cody Mactod. "Hi i'm here to do an inventory check for Tetron really quick," Cody said. So we have the Knife of Doom, Sword of Azak, and the key of job, all priceless weapon/artifacts. I'll be needing those. Now the login password is Tetron1. let me swap one billion pounds,

And five billion UDC. That should do the trick. Let's go goo! He yelled, pointing towards the backdoor left open. "I smell a thief," Mega-sphere said. "You Ultra-Bots never give up," Cody said. It's too late, the wires are already gone through. "no worries, your life will suffice," Mega-sphere said. "Mega-sphere to headquarters. We've got a trader on our hands. Looks like he's trying to run out. "Tetron, you want him dead or alive?" He asked. "Dead" replied Tetron. "Target presumably dead," replied Mega-sphere

CHAPTER 13

Home Base

Location: Bergall City

"Through this door," said Sarah. Wait, there's something here. "looks like an explosion," she said. I recognize her as the feathered one. Zeda, is that you? Asked Bruce. "Yes it's me and my leader maximus," Zeda said, lifting up maximus." He's wounded from the blast," she said. "Ok, we're returning to base," Sarah said. Our transports are right outside this city wall. "they can come with us. We have a medic on standby," she said. "Thank you, and your base's location is where, exactly?" asked Zeda. "Qubar!" Shouted Bruce.

CHAPTER 14

War

Location: Bergall City

"On this day the humans infiltrated bergall and stole from us," Tetron said, pacing back and forth. Let us remember our fallen comrades. "As we rebuild the city, we prepare for war!" yelled Tetron.

To be continued..

World renowned street author is back! This time to take the undisputed book world by storm. Follow him on his writer's journey as he delivers action packed storylines, and cinematographic viewpoints. All throughout amazing titles.

SIGNATURE

AI Writers & Publishing
DocuSign - Kerrick Patterson
Non Fungible Token - Front/Back Cover Artwork
NFT Owner - SR Animation

Book Publishing Contract

This Book Publishing Agreement (the "Agreement"), is entered into on this ___9/11/2023___ day of ___9/11/2023___, 20___ (the "Effective Date"), by and between ___KERRICK DEWAYNE PATTERSON___ ("Author"),

DOCUSIGN

and _____ ("Publisher"), a _____
organized in the state of ___Atlanta, Ga___ , whose principal office is located
at _____.

WHEREAS, the Author is or shall be the author of a literary work either tentatively or officially entitled
___Battleworld___

as defined in **Exhibit A** attached to this Agreement.

WHEREAS, the Author desires to have the Publisher publish the literary work and the Publisher desires to publish the literary upon the terms and conditions set forth herein.

NOW, THEREFORE, in consideration of the mutual covenants, terms, and conditions set forth herein, and for other good and valuable consideration, the receipt and adequacy of which are hereby acknowledged, the parties hereto agree as follows:

Author's Grant of Rights. Under the terms and conditions of this Agreement, the Author hereby grants the below-referenced permissions to the Publisher for the full term of the United States copyright in the literary work, including any renewals and extensions thereof, the exclusive right throughout the world to:

a. print, publish, and sell the literary work in print book form in the English language, unless otherwise stipulated, and in languages other than English in the following editions:

- ☒ Trade hardcover
- ☒ Trade paperback
- ☒ Mass market paperback
- ❏ Other:
  Audio Books

b. authorize on the Author's behalf the right to print, publish, and sell the literary work in the English language, unless otherwise stipulated, and in languages other than English:

- ☒ in trade or other hardcover, paperback, and large-type reprint editions
- ☒ as an audiobook
- ☒ As an electronic book (e-book)
- ❏ Other
  Comic Book etc

33

The aforementioned rights are referred to collectively herein as ("Subsidiary Rights"). The Publisher will provide the Author with copies of any Subsidiary Rights licenses granted to third parties under this Book Publishing Agreement. The Author may terminate the Publisher's authorization to license any Subsidiary Right at any time after _____ 10/14/2023 [months/years] after the first publication of the United States trade hardcover edition of the literary work in the English language, for any country and/or language for which

Use of Author's Name. The Publisher, and any licensees or assigns of the Publisher's rights under this Agreement, have the right to use the Author's name and any approved image, or the like, and biography for advertising, marketing, and promotion of the literary work and other rights permitted under this Agreement.

Author's Reservation of Rights. All rights not expressly granted by the Author to the Publisher under this Agreement are reserved by the Author. The Author shall promptly notify the Publisher of the intention, exercise, or disposition (whichever is earlier) of any right that is reserved by the Author.

Manuscript Delivery. The Author shall deliver Unlimited copies of the final and complete manuscript of the literary work, as well as an electronic file in _____ PDF _____ format containing the manuscript of the literary work no later than _____ 10/14/2023 _____ ("Manuscript Due Date").

 ❑ The Author shall also supply by the Manuscript Due Date, at the Author's expense, all photographs, drawings, charts, indexes, or other materials mutually agreed upon as necessary to the completion of the manuscript. The Publisher will consult with the Author within _____ days of receipt of the manuscript whether or not the literary work is in the proper form or if any revisions are required.

 ❑ The Author will have _____ days/weeks to make such changes or revisions required by the Publisher. Should the Author fail to deliver the revised Manuscript within the prescribed time, the Publisher may terminate this Agreement upon written notice to the Author.

In the event the author does not deliver the complete manuscript of the literary work by the Manuscript Due Date, the Publisher may terminate this Book Publishing Agreement upon written notice to the Author. The Publisher also reserves the right to charge the cost against any sums procured to the Author.

a If the Publisher terminates this Agreement all of the parties' obligations under this Agreement cease except for those that expressly survive this Agreement. Should this occur, the Author agrees to return any advances already received within _____ days to the Publisher. If, after termination for non-delivery the Author does complete the Work, the Author sha,, first offer the literary work directly to the Publisher for consideration for publication on the terms and conditions set out in this Book Publishing Agreement before publishing the literary work or

EXHIBIT A

Additional Services and Obligations: Nft Cover artwork

Other: Owned and Distributed by Kerrick Patterson. AI Writers and Publishing

IN WITNESS WHEREOF, the undersigned have executed this Book Publishing Agreement effective as of the _____14th_____ day of ___October___, 20 _23_ (the "Effective Date").

DOCUSIGN

Dated: _____   Dated: _____

DocuSigned by: DocuSigned by:

kERRICK DEWAYNE PATTERSON   kERRICK DEWAYNE PATTERSON
AD417E7C858742B   AD417E7C858742B
_____   _____
Publisher's Signature   Author's Signature

_____   _____
        AI Writers & Pub   KERRICK DEWAYNE PATTERSON

_____   _____
Publisher's Printed Name or Entity   Author's Printed Name

Publisher's Contact Information: **Author's Contact Information:**

Address: _____   Address: _____
_____   _____

Phone Number: _____   Phone Number: _____

Email Address: _____   Email Address: _____

Milton Keynes UK
Ingram Content Group UK Ltd.
UKHW051330181023
430741UK00031B/63

9 781088 278741